INGANEKWANE YETINOMBHOLO

THE NUMBER STORY

SMALL BOOK ONE

ENGLISH - SISWATI

*Numbers Teach Children
Their Number Names*

written and illustrated by

MISS ANNA

Early Reader Edition of *The Number Story 1*
Bronze Medal Winner, 2016 Wishing Shelf Book Award

Library of Congress Control Number: 2018902040

Names: Miss Anna, author.
Title: Number story : numbers teach children their number names / Miss Anna.
Description: Portland, OR: Lumpy Publishing, 2018.
Identifiers: ISBN 978-1-945977-93-0 | LCCN 2018902040
Summary: The pictures and rhymes present stories which introduce numbers 0-10.
Subjects: LCSH Numeration—English--SiSwati--Pictorial works--Juvenile literature. | BISAC JUVENILE NONFICTION /
Languages: English--SiSwati
Classification: LCC QA141.3 .M57 2018 | DDC 513—dc23

Publisher: Lumpy Publishing
Website: www.missannabooks.com
Email: missanna@missannabooks.com

Paperback: ISBN 978-1-945977-93-0
Printed in the U.S.A.   1 3 5 7 9 10 8 6 4 2

Ufuna kufundza emagama
Etinombholo tetfu?

It is very easy and a lot of fun!

Kumalula kakhulu kantsi futsi kumnandzi!

Say-along our little jingle

Hlabelela natsi indzatjana lencane!

starting from Number One!

Asicale enombholweni yekucala!

ONE looks like my one finger.

KUNYE

Kubukeka kunjengemuno wami.

ONE!
KUNYE!

2
TWO trails a tail.
KUBILI
Uzama umsila!

A TAIL! UMSILA!

3

THREE has bumps.

KUTSATFU

Inemabhampi.

BUMPY!

INEMABHAMPI!

4

FOUR carries a sail.

KUNE

Kutfwele liseyili.

LISEYILI!

Sikebhe lesine Liseyili!

5

FIVE   is a racing track.

SIHLANU

Ngumvila wekuncintisana.

VROOM
VROOOM!

6

SIX   curves like a snail.

SITFUPHA

Kulijika njengemnenke.

A SNAIL! UMNENKE!

7

BE CAREFUL! IT'S SHARP!
Caphela! Kucijile!

8

EIGHT is rollercoaster rails.

SIPHOHLONGO

Kusipolo se-*rollercoaster*.

YAY!
YIPPEE!

9

NINE   is a bubble on a stick.

IMFICA

Ligwebu endvukwini.

A BUBBLE!

LIGWEBU!

10

TEN is an eye of a whale.

LISHUMI

Liso le-whale.

WINK!

FICA LISO!

HELLO! SAWUBONA!

And
Na

0

ZERO   is an empty pail.

KUTE

Kwekukhelela lokungenalutfo.

IT'S EMPTY!
AKUNALUTFO!

Thank you for playing with us today.

We had a lot of fun too!

Siyabonga kudlala natsi namuhla.

Sibe nesikhatsi sekujabula natsi!

We are your Number friends,
Zero to Ten,
Who will be here for you~
Sibangani bakho bekucala
Kusuka lakungenalutfo
khona kuya eshumini.
Ngubani lotoba khona kuze abe nawe.

Bye-bye now!
See you again soon!
Sala kahle manje!
Sitokubona futsi masinyane!

The Numbers are *SINGING* too!

To sing-a-long, look for Miss Anna Number Story
at your favorite music store like iTUNES.

MP3

Numbers 0-10
IDENTIFYING
& COUNTING

Numbers 11-20
& Ordinals

first, second, third...

Numbers 0-100
& Place Values

ones. tens. hundreds...

About Clocks
& Telling Time

hours. minutes. seconds

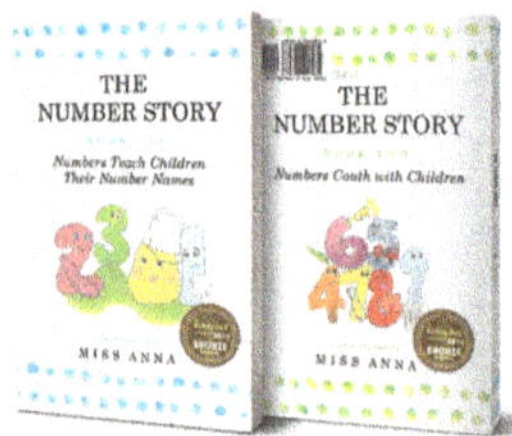

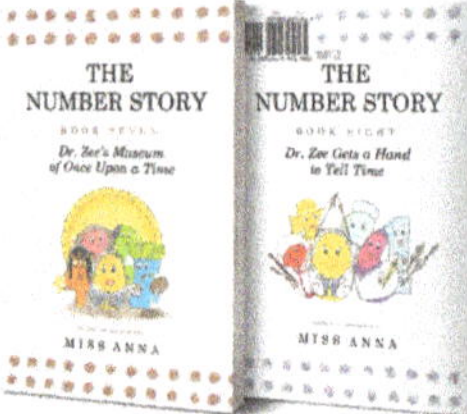

Number Story 1 & 2

isbn: 978-0-996216-48-7

Number Story 3 & 4

isbn: 978-1-945977-01-5

Number Story 5 & 6

isbn: 978-1-945977-06-0

Number Story 7 & 8

isbn: 978-1-949320-40-4

For more Miss Anna books to love,
visit us at

w w w . m i s s a n n a b o o k s . c o m

Numbers are working hard all over the world!
Come Travel the World with Us!